The Solar System:
God's Heavenly Handiwork

JASON LISLE

INSTITUTE FOR
CREATION
RESEARCH

Dallas, Texas
www.icr.org

As Director of Research, Dr. Lisle leads ICR's gifted team of scientists who continue to investigate and demonstrate the evidence for creation. He graduated *summa cum laude* from Ohio Wesleyan University where he double-majored in physics and astronomy and minored in mathematics. He earned a master's degree and a Ph.D. in astrophysics at the University of Colorado. Dr. Lisle specialized in solar astrophysics and has made a number of scientific discoveries regarding the solar photosphere and has contributed to the field of general relativity. After completion of his research at the University of Colorado, Dr. Lisle began working in full-time apologetics ministry, focusing on the defense of Genesis. Dr. Lisle was instrumental in developing the planetarium at the Creation Museum in Kentucky, writing and directing popular planetarium shows including "The Created Cosmos." Dr. Lisle speaks on topics relating to science and the defense of the Christian faith using logic and correct reasoning; he has authored numerous articles and books demonstrating that biblical creation is the only logical possibility for origins.

THE SOLAR SYSTEM: GOD'S HEAVENLY HANDIWORK

by Jason Lisle, Ph.D.

First printing: April 2014

All Scripture quotations are from the New King James Version.

ISBN: 978-1-935587-60-6

Please visit our website for other books and resources: www.icr.org

Printed in the United States of America.

Table of Contents

Introduction

One of the wonderful things about astronomy is that it is so different from our everyday experience. Things are not what they might seem at first glance. Who could have guessed that those tiny little specks of light in our night sky are actually "suns" hundreds of times larger than Earth? Who would have suspected that the "evening star" is actually a rocky planet about the same size as our own? How unexpected to find that the solid earth beneath our feet is actually moving at 67,000 miles per hour around the sun, all the while spinning like a top! God has constructed the universe in a truly marvelous way. As we study it, the universe continually surprises and delights us by challenging our understanding of how things work.

Every new discovery in astronomy is a delightful revelation that God is even more amazing, creative, and powerful than we previously supposed. In this booklet, we explore aspects of the solar system, including our own Earth. Through them we will see that truly "the firmament shows His handiwork" (Psalm 19:1).

The Solar System

Our solar system is a great example of God's handiwork. It consists of the sun and everything that orbits it, including the eight planets, asteroids, comets, centaurs, trans-Neptunian objects, and dust. The largest

and most massive object is the sun, a sphere of hydrogen and helium gas held together by its own gravity. With a diameter of 865,000 miles, the sun is 109 times wider than Earth and constitutes 99.8 percent of the solar system's mass. It appears so small in our sky because it lies an amazing distance away—93 million miles. A car travelling at 60 miles per hour would take 176 years to travel such a distance.

The Planets

The next largest objects are the planets. Jupiter is the largest, with a diameter of 86,881 miles. Saturn is next, then Uranus, Neptune, Earth, Venus, Mars, and Mercury. All eight planets orbit the sun in the same direction (counterclockwise as viewed from above Earth's north pole) and are very nearly in the same plane, which is called the *ecliptic*.

The four planets nearest the sun—Mercury, Venus, Earth, and Mars—are called *terrestrial* ("earthlike") and are relatively small worlds with dense, rocky compositions. The remaining four—Jupiter, Saturn, Uranus, and Neptune—are *gas giants* or *Jovians* ("like Jupiter"). Much larger than terrestrials, they are comprised primarily of hydrogen and helium gas. As with the sun, these balls of gas are held together by their own gravity. Uranus and Neptune are smaller than the others and are sometimes called *ice giants* due to their high abundance of various forms of ice.

Solar System Distances

Distances in the solar system are often listed in terms of astronomical units (AU). One AU is the average distance between Earth and the sun (about 93 million miles). The terrestrial worlds are all within 3 AU, but the gas giants orbit considerably farther out. Jupiter orbits at 5.2 AU, and Saturn is 9.54 AU—putting it at around 1 billion miles from the sun!

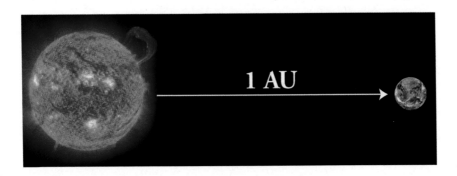

Uranus is 19.1 AU distant, and Neptune's distance is 30 AU—30 times farther from the sun than Earth.

Planetary Orbits

In the 17th century, creation scientist Johannes Kepler discovered that planets orbit in ellipses—"squashed" circles. An ellipse is the set of all points in a plane whose distance from two fixed points (called *foci*) gives the same sum. Kepler found that the sun was located exactly at one focus (the other focus is empty). The fact that planets orbit in ellipses with the sun at one focus is referred to as Kepler's first law of planetary motion.

Kepler also discovered that planets speed up when they are closer to the sun and slow down when farther away. This is Kepler's second law of planetary motion. He further found a relationship between the size of a planet's orbit and the time it takes the planet to go around the sun once. Specifically, the square of the period is equal to the cube of the distance from the sun. Planets that orbit close to the sun have short periods, whereas those that orbit far away have very long periods.

Why these laws work wasn't known until Isaac Newton's discoveries of the laws of motion and gravity allowed him to mathematically prove all three of Kepler's laws from first principles. Newton rigorously proved that gravity is the cause of the orbital motions of planets. The closer a planet is to the sun, the faster it orbits because the sun's gravity is stronger.

The Laws of the Universe

Bible critics sometimes view laws of nature as a replacement for God's power. But the Bible teaches that God directly controls the universe. God is not a god of confusion (1 Corinthians 14:33), but by the expression of His power upholds all things in a consistent and often predictable way (Hebrews 1:3). Laws of nature are not a substitute for God's power, but are rather examples of it. God's consistent and law-like sovereignty over the universe makes astronomy possible.

When we contemplate the sizes of these worlds, the distances involved, and the God who holds every atom in its place, it is amazing to think such a God would show so much compassion and mercy toward us. "When I consider Your heavens, the work of Your fingers, The moon and the stars, which You have ordained, What is man that You are mindful of him, And the son of man that You visit him?" (Psalm 8:3-4).

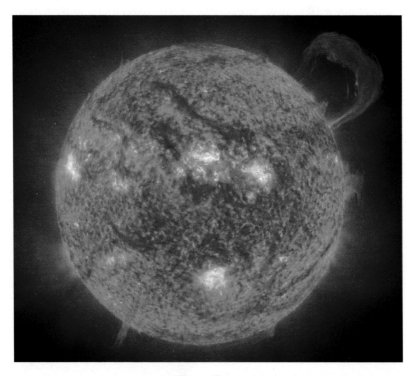

The Sun

The sun is remarkable in its complexity and power. It and the other luminaries were created on the fourth day of creation to separate day from night, help us mark the passage of time, and give light upon the earth (Genesis 1:14-15). A fourth purpose is to declare God's glory (Psalm 19:1-6).

The sun and moon are described as "great" lights. A fifth purpose of these two lights is to "govern" the day and the night (Genesis 1:16). The sun can be said to "rule" over the day because it defines the day and overpowers all other luminaries. The moon "governs" the night by outshining all other nighttime luminaries, though it is not always visible and the stars can "rule" in its absence (Psalm 136:9).

Properties of the Sun

The sun is 109 times the diameter of Earth and over a million times its volume. It comprises 99.8 percent of the solar system's mass. If a 10-pound bowling ball represented the sun's mass, then everything else in our solar system could be represented by the combined mass of one nickel and one penny. Jupiter would be the nickel.

The sun is made up almost entirely of hydrogen and helium gas and is so hot that for most of its interior, the atoms are completely ionized—their electrons have been stripped away from their nuclei.

Solar Structure

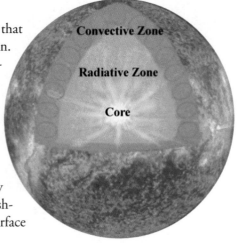

The sun is divided into layers that differ by temperature and motion. The core exceeds 15 million degrees Celsius. At such temperatures, the protons from the hydrogen atoms move so quickly that they smash into one another and form helium. This process, called *nuclear fusion*, releases an enormous amount of energy that propagates outward, replenishing the energy that the solar surface radiates into space.

The radiative zone extends outside the core to about two-thirds the radius of the sun. It is still millions of degrees but is not hot enough for nuclear fusion. In the outermost third—the convection zone—ionized gas moves in large overturning cells on multiple scales of incredible complexity.

The photosphere, the sun's visible surface, is about 6,000 degrees Celsius. The smallest overturning convection cells, called *granules*, are visible in high-resolution images. The photosphere often has small darker regions called *sunspots* caused by magnetic fields that inhibit convection, preventing energy transport from below.

Beyond the photosphere is the nearly transparent chromosphere, and beyond that is the solar corona—a large envelope of extremely thin and highly structured ionized gas that surrounds the visible disk of the sun. Paradoxically, the corona is much hotter than the region below it, with temperatures exceeding one million degrees Celsius. The exact mechanism by which the corona is heated is not precisely known.

Designed for Life

In many ways the sun is an ordinary star, but in others it is clearly designed for life to be possible on Earth. Some stars have super-flares that release enormous amounts of deadly radiation. The sun's flares are mild, and its temperature and distance from Earth are ideal for life. By contrast, hotter stars produce far more ultraviolet radiation that would have harmful effects on living tissue. And cooler stars emit far more infrared "heat" for a given amount of visible light.

Even the position of the sun in the galaxy seems optimized for life and for science. If it were close to the galactic core, harmful radiation could be a big problem. If it were on the outer rim, half the sky would be nearly void of stars, making it harder to measure seasons or investigate the universe.

The Sun Confirms Creation

Secularists believe that the sun has been fusing hydrogen for nearly five billion years. But nuclear fusion gradually changes the density in the core, causing a star to brighten over time. If the sun were billions of years old, it would have been 30 percent fainter in the distant past—Earth would have been a frozen wasteland and life would not have been possible.

Secular astronomers currently believe that the sun (as with other stars) was formed by the collapse of a nebula—a giant cloud of hydrogen and helium gas. Astronomers have discovered thousands of nebulae, but no one has ever seen one collapse in on itself to form a star. The outward force of gas pressure in a typical nebula far exceeds the meager inward pull of gravity. As far as we know, nebulae only expand and never contract to form stars. Even if gravity could somehow overcome gas pressure, magnetic fields and angular momentum would tend to resist any further collapse, preventing the sun from forming at all.

When we examine the science of the sun, we find that it confirms what Scripture teaches: God made the greater light to rule the day.

***

Mercury

Mercury is the smallest and least massive of the eight planets. This innermost planet is nearly three times closer to the sun than Earth. It is a solid, rocky world, with only a trace of an atmosphere. It has mountains, valleys, plains, and lots of craters. When it comes to creation research of the early solar system, Mercury provides many interesting clues.

A World of Extremes

Mercury takes only 88 Earth days to complete one orbit. The *sidereal* (relative to the stars) rotation rate of Mercury is 59 days. If we could watch Mercury through a telescope from a distant star, we would see it rotate one time every 59 days. But if we could watch Mercury from the sun, it would appear to turn once every 176 days. The reason for the dif-

ference is that Mercury orbits the sun as it rotates. This is also true of the other planets, though the difference is generally much less.

Because of Mercury's slow rotation, a given spot on the surface is in direct sunlight for about 88 Earth days. Since there is no substantial atmosphere to transport the heat, the surface temperature on the day side of Mercury can reach 800°F—hot enough to melt lead. And the night side of Mercury can drop to -280°F!

Mercury's Strange Orbit

Mercury's rotational (sidereal) period is precisely two-thirds of its orbital period, so Mercury turns on its axis three times every time it orbits the sun twice. Whenever the ratio of two periods can be expressed by a simple fraction, it is called a *resonance*. Mercury is the only planet in our solar system whose rotation period and orbital period are in resonance. The reason for this involves Mercury's orbit. Mercury has the most eccentric orbit of any planet—meaning that its orbit is noticeably elliptical and not as circular as the other planets. From Kepler's second law, a planet in an elliptical orbit moves faster when closer to the sun than it does when farther away. The point of closest approach is called *perihelion*, when Mercury moves the fastest. The most distant point is *aphelion*, when it orbits the slowest. But Mercury's rotation rate does not change. Interestingly, when Mercury is near perihelion, its rotation rate essentially matches its revolution rate, such that it keeps the same side pointed at the sun for several weeks.

Observing Mercury

From Earth, Mercury appears quite bright, but it can be a challenge to locate because it is easily lost in the sun's glare. Most other planets can be seen late at night, but not Mercury—it is only visible at twilight and only at certain times of the year when it is in a part of its orbit that appears (in angle) most distant from the sun. This position is called *greatest elongation*. At such times, it is possible to see Mercury just after sunset or just before sunrise. Thanks to Mercury's short period, greatest elongations happen six or seven times a year.

Mercury Confirms Creation

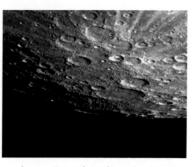

In 1974, the *Mariner 10* spacecraft was able to image much of the day side of Mercury in unprecedented detail. It also measured a substantial magnetic field. If our solar system is billions of years old, a small planet like Mercury should not be able to maintain a magnetic field for so long. But this doesn't surprise creationists.

Physicist Russell Humphreys has an interesting biblically based model that is able to account for the current strength of planetary magnetic fields based on their true age of about 6,000 years. His model lines up quite well with the current measured field strengths of the planets. Dr.

Humphreys predicted that Mercury's magnetic field would show measurable decay since the 1974 measurements, which was confirmed by the more recent *Messenger* spacecraft.

Creationists have long debated when and how the craters in our solar system actually happened. Were the planets created with craters? Did God use some process to make the planets such that the craters are the last material He brought to impact the surface? Did they happen after the Curse or during the Flood year? If so, why? Most other rocky worlds in our solar system have tectonic, atmospheric, and volcanic activity that can remove evidence of any previous craters—but not Mercury. Its pristine surface may be a window into the original conditions of our solar system. What other secrets can we discern from this fascinating little world? Time will tell.

Venus

For millennia, people have enjoyed the sight of the "evening star" shining brightly in the western sky or the "morning star" appearing before sunrise in the eastern sky. However, Venus is not a star at all but a small rocky planet about the same size as Earth. Its composition is also similar to Earth, and its orbit is physically closest to Earth's of all the planets. For these reasons Venus is sometimes referred to as Earth's sister or Earth's twin. But there are far more differences than similarities.

Properties of Venus

Dense clouds of sulfuric acid and toxic sulfur dioxide mask all surface features on Venus. The atmosphere, comprised mainly of carbon dioxide, is the thickest of the four terrestrial planets. Since Venus orbits 26 million miles closer to the sun, it endures nearly twice the solar energy as Earth. The atmosphere traps this energy and causes the surface temperature to approach 900°F—the hottest of any solar system planet.

In 1990, NASA's spacecraft *Magellan* began systematically mapping the surface using radar, which is not blocked by clouds. *Magellan* revealed many of the same geological features as Earth, including mountains, valleys, canyons, volcanoes, lava flows, craters, and plains, along with two major "continents"—Ishtar Terra and Aphrodite Terra.

Venus also has different geological features, such as arachnoids, novae, tesserae, coronae, and pancake domes. Arachnoids are regions where the terrain has been folded and broken into a gossamer-looking structure resembling a spider web. Novae are "radially fractured centers" between 60 and 200 miles in diameter. Tesserae are complex, ridged terrains found on plateaus. Coronae are oval-shape features thought to be produced by plumes of upwelling magma that cause the surface to bulge and then collapse in the center, forming a crown-like ring. Pancake domes are similar to shield volcanoes on Earth but are flatter and broader. Venus has no moons and virtually no magnetic field.

Earth and Venus are nearly identical in size and bulk composition and have similar orbits. In the secular view, they have a similar history too. So why would Earth's "sister" be so radically different from Earth? There are secular speculations for such things—but such diversity is expected in the Christian worldview.

A Day on Venus

Venus has the most circular orbit of any solar system planet. Its axial tilt is only three degrees, so there are no seasons. Since it is closer to the sun than Earth is, Venus orbits faster and completes a circuit every 7.4 months. But since Venus rotates once every eight months, its day is actually longer than its year. This is the sidereal day—the rotation of Venus relative to the stars. All eight planets orbit the sun counterclockwise, as viewed from the solar system's north pole. Most of the planets also rotate counterclockwise, but Venus rotates *backward*. On Venus, the sun would rise in the west and set in the east.

Secularists do not have a good explanation for this. They think the solar system formed from the collapse of a rotating nebula. The natural expectation would be that all planets would rotate in the same direction at about the same rate, and they would all have very little axial tilt. Venus rotates exactly the opposite of what the evolutionary models require. But the biblical view expects such diversity.

The backward rotation of Venus causes its solar day to be much *shorter* than its sidereal day—a unique phenomenon in the solar system. Recall that the solar day is the average time from one sunrise to the next as viewed from a planet's surface. This is different (and normally slightly longer) from the sidereal day because planets orbit the sun and not the stars. Since Venus rotates in the opposite direction, its solar day is reduced to 3.8 months.

Phases

Venus is not uniformly illuminated but appears bright on one side and dark on the other. Just like the moon, Venus goes through new, crescent, quarter, gibbous, and full phases as seen from Earth. But there are three differences in the phases of Venus compared to the moon.

First, the moon takes roughly one month to go through its phases, whereas Venus takes over 19 months. Second, whereas the moon appears about the same size as it goes through its phases, Venus does not. In its gibbous phase (nearly full) it is on the far side of the sun and appears very small, while in its crescent phase it appears very large because it is nearly as close to Earth as it gets.

Finally, the phases of Venus (and Mercury) are reversed relative to the moon. Both Venus and the moon orbit in the same direction, so how can this be? The difference is because Venus orbits the sun, whereas the moon orbits Earth. So when the moon is in between Earth and the sun, it is moving to the left (eastward). But when Venus is in between the sun and Earth, it is moving to the right (westward).

From Earth, Venus stands out as a pure white light, superior in splendor and luminance. Venus is mentioned in Scripture as the "morning star," where its brilliance is used as a symbol for Christ (Revelation 2:28; 22:16). None of the other nighttime stars can compete with Venus, so it is a fitting symbol of the beauty and glory of our Lord.

Earth and Moon

From a secular perspective, Earth is just a tiny bit of rock and water in a vast and meaningless universe of chance. But in the Christian worldview, this pale blue world is the most important planet in the universe.

Properties of Earth

Earth orbits the sun at an average distance of 93 million miles and completes an orbit in one year. Earth's solar day is 24 hours, and it takes 23 hours and 56 minutes to rotate once relative to the stars—a sidereal day.

Earth's physical properties are similar to those of the other terrestrial worlds orbiting close to the sun. They all have mountains, valleys, rifts, canyons, and craters. Earth is the largest in diameter but the sizes are not all that different. Despite these similarities, Earth is unique in many ways.

Uniqueness of Earth

Earth is the only planet known to contain life. Creatures flourish in virtually every environment, in striking contrast to the other planets' lifeless surfaces. Many of Earth's unique qualities seem specifically designed to support such life.

Earth's distance from the sun and atmospheric pressure are just right for liquid water, which covers over 70 percent of its surface. No other known planet has such an abundance. Since water is essential to life, its presence here seems to be a key design feature.

Earth's atmosphere has a protective ozone layer that partially blocks ultraviolet radiation. Radiation can be very damaging to living tissue, so this too is a design feature. Unlike Venus, Earth has a strong magnetic field that deflects harmful cosmic radiation. Its strength is consistent with Earth's biblical age of around 6,000 years but is wildly inconsistent with the secular assumption of billions of years.

Earth is tilted on its axis 23.4 degrees relative to its orbit, causing it to experience seasons. This tilt appears to be well-designed for life. If Earth were tilted less, the polar regions would receive less energy, reducing the habitable area of the planet. If the earth were tilted more, the seasons would become more extreme, potentially reducing plant-growing seasons and making the environment less hospitable.

The Moon

Earth's moon is over one quarter the size of Earth in diameter. No other planet has a moon this proportionally large. The moon aids life

on Earth by inducing tides, which clean the shorelines and prevent the oceans from stagnating. It also provides light at night.

The lunar surface is barren and rocky, with heavily cratered highlands. It also has lower, relatively smooth regions called *maria* (Latin for "seas") that appear as large dark regions. Apparently, they are large impact basins that have filled in with magma, erasing any previous record of cratering. Without an atmosphere to redistribute thermal energy, the temperature on the moon can exceed 200°F during the day and drop to -280°F at night.

The moon takes 27.3 days to rotate once, which is exactly how long it takes to orbit Earth. For this reason, observers on Earth only see one side of the moon. The fact that the rotation and revolution of the moon have exactly the same period is called tidal locking. All large and many small moons in our solar system are tidally locked.

The Uniqueness of the Moon

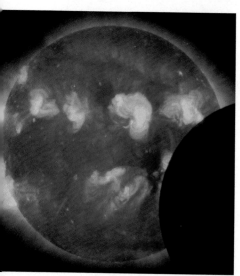

The moon is both 400 times smaller and 400 times closer to Earth than the sun. The moon and sun have about the same apparent size in our sky on average, making total solar eclipses possible. Earth is the only known planet that can experience eclipses where its moon so precisely covers the sun.

The moon orbits very close to the ecliptic—the plane of Earth's orbit around the sun. This makes solar and lunar eclipses more common on Earth than they would be if the moon orbited around the planet's equator as most other moons do. Yet, because the moon does not orbit *exactly* in the ecliptic, we do not have eclipses every month.

A Young Moon

The moon moves away from the earth about 1.5 inches every year—a process called recession. (The rate would have been faster in the past since the moon would have been closer.) If we extrapolate this effect, we find that the moon would have been touching Earth 1.4 billion years ago, which is inconsistent with the secular age estimate of four billion years for Earth and the moon. The moon would have been only 730 feet closer to Earth 6,000 years ago, so lunar recession is not a problem for the biblical timescale.

Conclusion

Earth is uniquely designed for life, and the moon is uniquely designed to aid that life. God chose to spend five of the six creation days making Earth just the way He wanted it to be. It is as if God took extra care to create it. Of all the planets in the universe, Earth is where God chose to place the creatures whom He made in His own image. It is our planet where Almighty God, out of His great love for us, took on human nature, died our death, and rose in glory.

Mars

Mars has features strikingly comparable to Earth's, with mountains, valleys, canyons, volcanoes, and polar ice caps. It even has some similar weather, including seasons, clouds, fog, wind, dust storms, dust devils, and occasional frost. Although liquid water is not found in any abundance, scientists have discovered substantial quantities of water-ice near

the poles and water vapor in the Martian atmosphere. Even the axial tilt and rotational period of Mars is much the same as Earth's.

Mars takes 24 hours and 37 minutes to rotate once on its axis—almost identical to Earth. As a result of its larger orbit, the Martian year equals 1.9 Earth years. Its thin atmosphere is composed mainly of carbon dioxide, and its force of gravity is only 38 percent of Earth's.

Martian Seasons

Like Earth, Mars' seasons result from axial tilt (25.2 degrees), *not* its elliptical orbit. However, Mars' orbit is significantly more elliptical than Earth's, which causes its distance from the sun to change and affects the *severity* of its seasons. So, even though, like Earth, Mars is closer to the sun during its northern hemisphere winter and farther away during its northern hemisphere summer, the effects are different. Its greater distance to the sun partially compensates for the increased duration and direct angle of sunlight experienced in northern hemisphere summers. And Mars' elliptical orbit causes seasons to be *less* extreme in its northern hemisphere than in its southern hemisphere.

In addition, Mars has polar ice caps that grow during the winter in their respective hemispheres and shrink during the summer—just like Earth's ice caps. But Earth's are water-ice, and Mars' ice caps are mostly water-ice layered underneath several feet of frozen carbon dioxide (dry ice).

Martian Topography

Most of Mars' surface resembles Earth's deserts, with rocks as far as the eye can see and very little relief. Though there are hills and even enormous mountains, they have gentle slopes that make them seem less magnificent than Earth's peaks.

Remnants of an ancient streambed on Mars

24

Several immense volcanoes exist on Mars, but most astronomers believe they are extinct and that Mars currently has essentially no geologic activity.

The surface of Mars has dry river beds and deltas. Evidence clearly suggests that Mars once had surface water. This is especially perplexing in light of the planet's thin atmosphere. Water can only exist as a liquid between certain temperatures and under sufficient atmospheric pressures, and the atmosphere of Mars is far too thin to allow water to be liquid for any length of time at any temperature. This mystery remains unsolved.

Martian Moons

Mars' two moons are quite tiny compared to Earth's moon. Phobos, the larger, is only about 10 miles across. Deimos has a width of only eight miles. Neither moon is spherical. This is common with small moons and asteroids since their gravity is insufficient to overcome the chemical bonds that prevent these bodies from collapsing into a spherical shape.

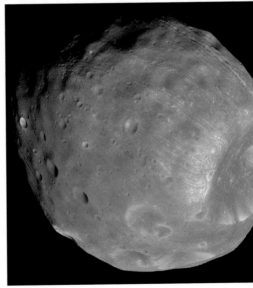

Phobos orbits Mars at an unbelievably close distance of only 3,700 miles—closer than any other moon to its planet. Its proximity to Mars—combined with Mars' gravity—means that Phobos orbits very quickly, with one orbit in only 7 hours and 39 minutes. At a greater distance, Deimos takes just over 30 hours to complete one orbit.

The origin of these moons is perplexing from a secular perspective. Were they asteroids

Phobos

captured by Mars' gravity? This is possible but involves an improbable chain of events. Moreover, captured asteroids are expected to have exaggerated, elliptical orbits, but Mars' moons orbit in nearly perfect circles.

As with so many aspects of the universe, the creative diversity of the Lord seems the best explanation for this puzzle. While posing a challenge for natural processes, the creation of unique moons in well-designed orbits is no problem for God.

Martian Opposition

Outer planets are best viewed when Earth passes between them and the sun. During such a configuration, the outer planet is said to be in *opposition* because it is opposite the sun. Most outer planets appear large and bright even when they are not in opposition, but because Mars is so small, the planet only looks bright (and large in a telescope) for a month or so around opposition. Unfortunately, because its orbital period is nearly twice as long as Earth's, Mars' opposition only happens about once every 2.1 years.

Mars and Earth possess great similarities but also vast differences. This is yet one more mark of the creativity of the Trinitarian God of Scripture. God Himself (Father, Son, and Holy Spirit) embodies a multitude of characteristics—diverse and yet unified. In the same way, the planets have unique variations representing the all-encompassing, endless ingenuity that the Creator exemplifies in all His forms. Indeed, the evidence of Him is clearly seen by what He has made—"even His eternal power and Godhead" (Romans 1:20).

Jupiter

Jupiter does not possess a solid surface but is an enormous spheroid of gas held together by its own gravity. It is composed primarily of hydrogen and helium, the same gases as the sun. However, its much cooler temperature allows the formation of molecules such as ammonia, water, and methane from various trace elements. These molecular compounds create Jupiter's colorful cloud formations.

Jupiter is massive—the equivalent of 318 Earths! If we could put it on a scale, it would weigh more than twice as much as all the other planets *combined*. Jupiter is so massive that its gravity slightly, but noticeably, affects the motions of the other planets.

A number of Jupiter's properties are difficult to explain in a multibillion-year framework. One is its strong magnetic field. Since magnetic fields naturally decay with time, how could Jupiter maintain such a powerful field over *billions* of years? And Jupiter emits nearly twice the amount of energy that it receives from the sun. If Jupiter were really billions of years old, why hasn't it cooled off by now?

Views of Jupiter

Powerful winds stretch clouds into colorful *belts* and *zones* that encircle Jupiter parallel to its equator. Belts are the dark brownish-orange features, and zones are lighter in color. Like Earth's jet streams, these clouds mark the differential rotation of Jupiter's atmosphere. The cloud formations are dynamic, growing or shrinking slightly from year to year and experiencing subtle changes in color.

The Great Red Spot, a large, somewhat permanent feature just below Jupiter's Southern Equatorial Belt, is of particular interest. It is a storm vortex roughly twice the width of Earth. Hurricanes on Earth are powered by warm ocean water and dissipate when they move over land. Since Jupiter has no land, its Great Red Spot continues indefinitely.

Jupiter rotates in less than 10 hours—faster than any other planet. This rapid rotation, combined with the extraordinary size of Jupiter, causes the planet to bulge at the equator.

The Galilean Satellites

When viewed through a telescope, four beautiful, bright moons appear in a nearly straight line. This is because they orbit in the plane of Jupiter's equator, and we see the Jupiter system from nearly edge-on. They were the first moons discovered orbiting another planet and are called the Galilean satellites in honor of their discoverer, Galileo.

These satellites are Io, Europa, Ganymede, and Callisto. Unlike gaseous Jupiter, these moons are solid bodies composed of rock and ice. Jupiter and its moons are like a miniature solar system with the inner moons orbiting faster than the others.

The Galilean moons are comparable in size to Earth's moon. Ganymede is the largest moon in the solar system at 50 percent larger than Earth's moon—it's even larger than the planet Mercury! If this moon orbited the sun directly, it would certainly be classified as a planet.

Io

More Moons and Rings

At latest count, Jupiter has 67 known moons—more than any other planet! Aside from the Galilean satellites, all these moons are very small, generally only a few miles across.

Based on their orbital properties, these moons fall into eight natural groups. One group consists of the only four moons that are closer than the Galilean satellites to Jupiter. The orbits of these inner moons and the orbits of the Galilean moons are nearly circular, are in the same plane as Jupiter's equator, and are *prograde*, meaning they orbit in the same direction that Jupiter rotates. The next seven beyond Callisto are also prograde, but their orbits tend to be tilted substantially, relative to Jupiter's equator, and are more elliptical than those of the inner moons. The remaining 52 moons also have tilted, elliptical orbits, but, amazingly, they are all *retrograde*—orbiting in the opposite direction of Jupiter's rotation. So, the region where moons orbit in one direction is separate from the region where they orbit the other way. This reduces the possibility of collision or close gravitational perturbations.

Jupiter also has a system of rings like Saturn. These rings are much less substantial than Saturn's, however, and the main rings consist of dust particles orbiting mostly around and inside the orbit of Jupiter's innermost moons, Metis and Adrastea. An even thinner "gossamer ring" extends farther out.

Conclusions

Most of what we now know about this fascinating planet was hidden from humanity for thousands of years. To our distant ancestors, Jupiter was a bright point of light in our night sky. Who would have imagined that this bright "star" would have so many remarkable characteristics? And what other celestial secrets has the Lord hidden for us to find and cherish?

Saturn

A rich system of rings composed of trillions of tiny moonlets—particles of water ice—orbits Saturn. Like Jupiter, Saturn is made of hydrogen and helium gas and trace amounts of molecules such as methane and ammonia. It also has colorful clouds that stretch into belts (dark-colored) and zones (light-colored), though its belts and zones are more subtle.

Nine times the size of Earth's diameter, Saturn has the lowest density of any planet and, amazingly, would actually float in water. At 890 million miles from the sun, Saturn takes 29.5 years to complete just one orbit.

Powerful storms occasionally develop and appear as bright regions within the belts and zones. Although Saturn has no permanent storms like Jupiter's Great Red Spot, its intermittent storms can last for many months. A Great White Spot manifests roughly every 30 years during the planet's northern hemisphere summer and can periodically occur in off-years as well.

Lord of the Rings

Though Jupiter, Uranus, and Neptune have ring systems, Saturn's rings are by far the most impressive. The main rings span 170,000 miles across but are less than one mile thick. The rings orbit in the plane of Saturn's equator. Fortunately, the planet's rotation axis is tilted 26.7 degrees in relation to the planetary orbital plane. Without this tilt, the rings would always appear edge-on, making them virtually invisible to us.

Saturn has several systems of rings that differ in brightness. The three major rings are the A-, B-, and C-rings. A is the outermost, B is in the middle and is the brightest, and C is the faintest. A small but noticeable void, the *Cassini Division*, occurs between the A- and B-rings. Other smaller divides also exist, such as the *Encke Gap* within the A-ring.

The spacecraft *Pioneer 11*, *Voyager 1*, *Voyager 2*, and *Cassini* have provided spectacular close-up views of Saturn, its rings, and its moons. *Pioneer 11* first detected the F-ring, a narrow filament of icy particles that orbits just beyond the A-ring and contains multiple threads twisted like the strands of a rope.

The Moons

Most of Saturn's 62 known moons are just a few miles across. Titan is the largest at 3,200 miles in diameter. It is the second-largest moon in the solar system (behind Jupiter's Ganymede). Like Ganymede, Titan is larger than the planet Mercury and would be classified as a planet if it orbited the sun directly.

Saturn and Titan

Titan is the only known moon with a thick atmosphere, composed primarily of nitrogen with traces of methane and other hydrocarbons. The existence of methane is a mystery for those who believe this moon is billions of years old; ultraviolet radiation from the sun is expected to break down methane in only a few tens of millions of years.

Mimas orbits close to the planet and has an enormous crater on one side that gives the moon a very strange appearance—a bit like the Death

Star from *Star Wars*. In contrast, Enceladus is white as snow and has only small craters. The *Cassini* spacecraft discovered plumes of icy material being ejected like geysers near its south pole, indicating significant internal heat. If the solar system is billions of years old, that heat should have escaped long ago. Enceladus does not experience enough gravitational tugging to regenerate its internal heat.

Saturn is the only planet known to have Trojan moons—moons that share a common orbit at precisely the same speed and thus never collide. Tethys shares its orbit with Telesto and Calypso, and Dione shares its orbit with Helene and Polydeuces.

Saturn's moons are partly responsible for the complex structure of its rings. Small gravitational fields of these moons can perturb ring particles into different orbits. Mimas' outside gravitational influence is responsible for the relative absence of moonlets in the Cassini Division. And Pan orbits within the Encke Gap and gravitationally deflects any moonlets that wander too close to it. Prometheus orbits just inside the thin F-ring, and Pandora orbits just outside. These *shepherd moons* gravitationally deflect any wayward moonlets back into the F-ring.

Janus and Epimetheus presented a perplexing riddle. Epimetheus is approximately 72 miles in diameter, and Janus is 111 miles. Both have nearly circular orbits around Saturn's equator. If the two moons were placed at the same starting line, Epimetheus would eventually lap Janus in about four years. The difference between the orbits of these moons is only 31 miles—*less* than the diameter of either moon. So how can they pass without collision?

As Epimetheus approaches Janus from behind, the gravity of Janus pulls forward on it—giving Epimetheus extra energy that causes it to move into a higher orbit. Meanwhile, Epimetheus' gravity pulls backward on Janus, causing Janus to lose energy and drop into a lower orbit. The two moons avoid collision by swapping orbits! Four years later, mutual gravity causes them to switch back into their original orbits.

What an amazing solar system the Lord has created! Almost every day some new scientific discovery gives us a little glimpse into the mind of God, and Saturn is certainly a wonderful example of His creative genius.

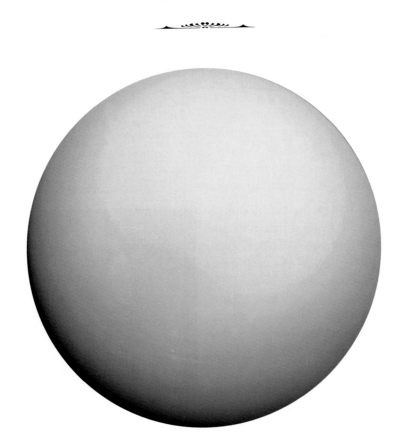

Uranus

As telescopes grew in size and optical quality, astronomers could investigate astronomical phenomena that are difficult or impossible to see with the unaided eye. In 1781, Sir William Herschel observed a small, light-blue disk. He initially supposed it to be a comet, but as comets generally have highly elliptical orbits and the extrapolated orbit of this blue disk was nearly circular, it had to be a planet.

Some favored naming the planet after Herschel, but tradition ultimately prevailed. The other five known planets (excluding Earth) were named after Roman gods. The new planet was named after the Greek god of the sky—Uranus (YOOR-un-us). Since Uranus is sky-blue in color, the choice seemed fitting.

Properties of Uranus

Uranus orbits the sun at an average distance of 1.79 billion miles and takes 84 years to complete one orbit. Its outer composition is mostly hydrogen and helium gas, with a small percentage of methane. Based on its density, its interior is thought to be composed of various ices such as water, ammonia, and methane. For this reason, Uranus is sometimes referred to as an *ice giant* rather than a gas giant like Jupiter or Saturn.

The *Voyager 2* spacecraft provided the most detailed images to date when it flew past Uranus in 1986, generating pictures of a nearly featureless blue sphere without the prominent belts and zones found on Jupiter and Saturn. Uranus occasionally manifests white clouds that are detectable in large, Earth-based telescopes.

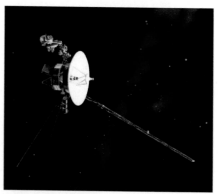

Uranus has a system of rings that are quite different from Saturn's rings. Saturn's main rings are broad sheets of orbiting material, while Uranus' rings are more like a series of 13 thin ropes. Each rope encircles Uranus at a discrete distance, and all are in the plane of its equator.

Unlike any other planet, Uranus rotates on its side. That is, the rotation axis is tilted approximately 90 degrees relative to the planet's orbital plane. The extreme tilt of Uranus is contrary to the expectations of the secular solar system model, in which the planets ought to have formed with their rotation axis nearly perpendicular to their orbital plane. Only Jupiter and Mercury meet this expectation. Secular scientists usually attribute the discrepancy to giant impacts in the distant past that knocked the planets from their original vertical orientations.

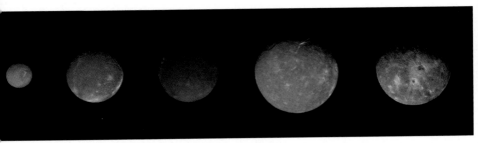

Left to right: Miranda, Ariel, Umbriel, Titania, and Oberon

Moons

Uranus has 27 known moons. The largest and brightest, Oberon and Titania, were discovered by Herschel in 1787 and are less than half the diameter of Earth's moon. Ariel and Umbriel were discovered in 1851, and little Miranda was discovered in 1948.

The remaining 22 moons are much smaller, all less than about 50 miles in radius and generally non-spherical. They were discovered during or after the *Voyager 2* flyby. In a new tradition, they were named after Shakespearian characters (mostly from *The Tempest*) or characters from Alexander Pope's poetry.

Uranus' moons are composed of various combinations of rock and ice. The 18 inner moons orbit in nearly perfect circles and in the planet's equatorial plane. They also orbit prograde—in the same direction Uranus rotates. Beyond Oberon, there is a considerable gap before the remaining nine moons. The outer moon orbits are not in Uranus' orbital plane, and each has its own unique orbital plane. Eight of the nine have retrograde orbits—opposite Uranus' rotational direction. This pattern of regular, prograde, coplanar moons being close to the planet and irregular, non-coplanar moons being at a greater distance seems to be a common feature of planets in the solar system.

Magnetic Riddles

The magnetic field of most planets is approximately aligned with their rotation axis. Not Uranus—its magnetic axis is offset from the rotation axis by an astonishing 60 degrees. Moreover, the magnetic axis does not pass through the center of the planet but is offset to one side by roughly one-third the planet's radius. From a secular perspective it is mystifying that Uranus has a magnetic field at all since magnetic fields should be nonexistent in planets that are billions of years old.

In 1984, creation physicist Russell Humphreys predicted the magnetic field of Uranus based on the amount of magnetic decay in the 6,000 years since its creation. *Voyager 2* confirmed this prediction. Although the presence of a strong magnetic field on any planet is a confirmation of recent creation, this is especially the case for Uranus.

To salvage their belief in billions of years, secular scientists have proposed that mechanical motion due to heat in a planet's interior somehow forms a dynamo, thereby regenerating the planet's magnetic field. But of the four giant planets in our solar system, Uranus alone lacks any measureable internal heat. So, there is no power source for the dynamo. Also, dynamo models predict that the magnetic field axis must be fairly well-aligned with the rotation axis. Uranus violates this condition as well.

The first planet to be discovered in modern times, Uranus is a diverse world of unique splendor. The properties of Uranus are fascinating and confound the secularists but clearly declare God's glory (Psalm 19:1).

Neptune

Neptune was known to exist *before* it was visually discovered. Creation scientist Isaac Newton mathematically demonstrated that the motion of planets could be explained by the sun's force of gravity deflecting their momentum into an elliptical path. This allowed astronomers to refine their calculations of planetary orbits to include the gravitational influence of other planets. The new physics correctly predicted the precise position for every planet...except Uranus.

French mathematician Urbain Le Verrier began considering whether another planet was pulling on Uranus. He mathematically computed the

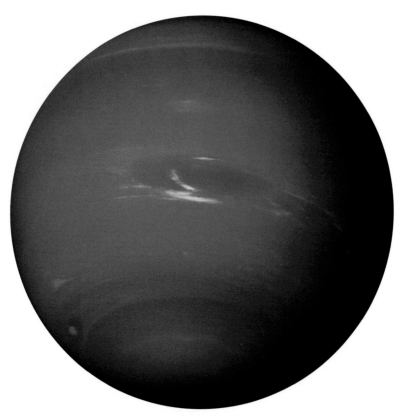

position that the unknown planet must have in order to explain the discrepancies. Le Verrier mailed his findings to Johann Galle of the Berlin Observatory, who received the letter on September 23, 1846, and, with the help of Heinrich d'Arrest, visually located Neptune that very evening.

Le Verrier christened the new world "Neptune" after the Roman god of the sea. The name fits the sea-blue color of the planet and indirectly pays homage to Isaac Newton by sharing the first two letters of his name.

Properties of Neptune

At an average solar distance of 2.8 billion miles, Neptune is the most distant planet of the solar system. It takes 164.8 years to orbit the sun. Physically, Neptune is a virtual twin of Uranus. Both worlds have an icy core surrounded by a thick atmosphere of hydrogen, helium, and small amounts of methane, which causes the blue color of both planets.

Neptune's largest moon is Triton, which is 23 percent smaller in diameter than Earth's moon. In contrast to all other large moons, Triton's orbit is retrograde—opposite the direction that the planet spins. Large moons generally orbit in the plane of their planet's equator, but Triton orbits at an angle of 23 degrees.

All of Neptune's other moons are much smaller and evaded detection for over a century. Nereid—just over 100 miles in diameter and with a highly eccentric (elliptical) orbit—was discovered in 1949. A third moon, Larissa, wasn't detected until 1981. The rest remained hidden until the *Voyager 2* encounter in 1989.

Triton

The Science of *Voyager 2*

One of the first discoveries of *Voyager 2* was the detection of a system of rings. At first, the rings appeared as arcs, only partially encircling the planet. But as *Voyager 2* drew closer, the rings were found to be complete, though thicker in certain places.

Neptune's five major rings seem to be a mosaic of the types of rings encircling the other Jovian (gas giant) planets. Three of them are thin threads, like the rings of Uranus; the two others are broad sheets, like Saturn's rings, but are thin like Jupiter's.

Voyager 2 discovered five small, new moons with circular, prograde orbits in the plane of Neptune's equator. With technological breakthroughs in ground-based imaging, several additional moons were discovered in later years, bringing the total known moons to 14.

The spacecraft also revealed a large dark spot in Neptune's southern hemisphere comparable in size to Earth and qualitatively similar to Jupiter's Great Red Spot. But, whereas Jupiter's red spot is relatively permanent, Neptune's dark spot was short-lived. In 1994, the Hubble Space Telescope revealed that the spot had disappeared and a new dark spot had formed in Neptune's northern hemisphere. It, too, was short-lived and has long since disappeared.

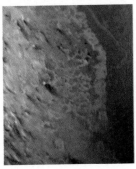

Voyager 2's images yielded another discovery—numerous horizontal, dark streaks in Triton's southern hemisphere. These were found to be "geysers" of nitrogen gas caused by solar heating of the frozen surface. Though the gas is transparent, the geysers pick up dark surface dust and launch it into Triton's tenuous nitrogen atmosphere. Eastward winds carry the dust many miles, accounting for the dark, horizontal streaks.

Confirmation of Creation

Unlike Uranus, Neptune has considerable internal heat, radiating more than twice the energy it receives from the sun. It is hard to imagine how such a process could last for billions of years. In addition, it is curious that Uranus lacks any internal heat, despite being nearly identical to Neptune. How can an evolutionary scenario make sense of this? Yet, this similarity-with-differences is a common characteristic that the Lord built into the universe. Diversity with unity is part of what makes science possible and is what we expect from the triune God.

Voyager 2 measured Neptune's magnetic field and found it to be similar in strength to that of Uranus, far stronger than we would expect if the planet were billions of years old. As with Uranus, Neptune's magnetic field is not even remotely aligned with the rotation axis and does not pass through the center of the planet. Such facts are difficult to account for in secular dynamo models but are consistent with the creative diversity of our Lord.

Pluto

Pluto was considered the ninth planet for 76 years, but in 2006 the International Astronomical Union voted to reclassify Pluto as a "dwarf planet." What prompted this change? Pluto is a small rocky/icy body at tremendous distance. Unlike any other planet, it showed no discernible size in even the largest telescopes of the period it was discovered. This meant that it had to be much smaller than Uranus or Neptune and no bigger than Earth.

As technology grew, the estimated size of Pluto shrank. We now know that Pluto is only 1,430 miles in diameter—about 18 percent the diameter of Earth—and has less than one percent the mass of Earth.

The new estimate of its small size is one reason for Pluto's demotion, but it's not the only reason. In 1992, astronomers detected another object orbiting beyond Neptune that was a mere 100 miles in diameter—far too small for a planet. The following year, five similar objects were discovered orbiting beyond Neptune. We now know of hundreds, a few of them nearly as large as Pluto. With these discoveries, astronomers began to